AF875228

Run-Off Leaves!

Its adhesive and camouflage properties create a striking impact on the ecosystem.

Jean-Paul Kamand

Copyright © 2025 by Jean-Paul Kamand.

All rights reserved.

No portion of this book may be reproduced in any form without written permission from the publisher or author except as permitted by copyright law.

Moses Filter Drain Team.

Individuals With All Abilities (IWAA).

Drain Team Literature.

Report online publication: 2003 - Revised & Updated Paperback Publication: 2025.

***ISBN:** 978-1- 7382643- 2- 2*

Dedication

In Love, in honour and memory of my Brother, ***Hugues Nawej***

Table of Contents

A Brief Background of Moses Filter and The Drain Team

Moses Filter is a reusable catch basin filtration system that offers a practical solution to reduce unintended urban runoff and advance environmental adaptability. Almost 100% of the runoff in city catch basins—including harmful discarded and dumped items—is carried through pipes into marine ecosystems.

© The Moses Filter. *Moses Filter is a reusable catch basin filtration system that offers a practical solution to reduce unintended urban runoff and advance environmental adaptability.*

The Moses Filter can be custom-made to fit municipal catch basins and operates to capture waste after it enters the catch basin and before it enters the aquatic ecosystem. We provide end-to-end services to custom-made appropriate filters, clean out the filters, sort the unintended waste, and dispose of it safely. We also supply downstream data analysis for municipality properties to apply to their upstream priorities. Moses Filter can be serviced through tailored employment initiatives for Individuals With All

Abilities, and it creates a viable, affordable option for socially and financially responsible storm drain maintenance that advances social and environmental goals.

It is with great honour that I recognize the management and staff at Coast Mental Health (Landscape with Hearts). The members and staff contribute significantly through their invaluable input and services, ensuring seamless operation during regularly scheduled cleanouts. I express my sincere gratitude for their diligent efforts and unwavering dedication, particularly for their presence during inclement weather, such as rainy, windy, and snowy days, consistently demonstrating their willingness to assist.

Abstract:

This study investigated the significant impact of leaf litter travelling downstream on cement or concrete roads toward municipal catch basins. With its adhesive characteristics, the leaf litter potentially carries billions of microplastics to our local marine ecosystem, which is a grave concern. It serves as a significant transport vehicle, hidden and undetected. The Moses Filter, a reusable catch basin filtration system, played a crucial role in our research. Our observations, collections, and analysis of leaf litter, facilitated by the Moses Filter, have provided vital insights into how microplastic waste may float undetected into our marine ecosystem. Our pilot testing occurred throughout the calendar year, utilizing the four seasons (Fall, Winter, Spring and Summer); however, we devoted our attention to the heavy rainfall months in the Pacific Northwest, which includes Vancouver, British Columbia, with US states Washington and Oregon (the fall and winter). Our research has potentially revealed a correlation between micro and nano-plastics in our bodies of water and leaf litter from urban areas. During heavy rainfall, water is forced through the trapped mainly leaves, causing the discharge of some pollutants and floatables attached to the leaves and ultimately ending in the local marine ecosystem. This study aimed to investigate the potential impact of leaf litter on the transportation of microplastics into municipal catch basins and local aquatic ecosystems, highlighting the urgent need for further research and immediate action. The relevance and importance of this topic cannot be overstated.

Environmental Science - Climate Adaptation

Environmental science is a fascinating field that continually fosters curiosity, beginning early in grade school. Students are introduced at a young age to the importance of observing our natural world: water, trees, leaves, and the weather cycle. 'Environmental Science represents a unique way of learning about our world. Alongside understanding the physical world and its phenomena and processes (Molnar et al., 2015), the fundamentals of scientific research methods have remained consistent, extending from grade school into adulthood.

As technology progresses at an accelerated pace, artificial intelligence (AI) is becoming increasingly ubiquitous. Is it possible to depend on the principles of science for consistency? Chiyangwa asserts, "Mobile and Internet-based technologies evolve swiftly, and the introduction of new hardware or software typically engenders corresponding innovations" (2015). We can rely on science and its foundational principles to maintain consistency and inclusivity. Science is grounded in "what can be observed, heard, and touched rather than in personal opinions or speculative conjectures" (Chalmers, 2013).

Science is an excellent vehicle for research as it fosters inquiry and offers hands-on opportunities and collaborative experiences that naturally engage students, communities, and policy decision-makers (Partridge, 2006). Science classes introduce us, as grade schoolers, to biology, often involving the dissection of a mammal to enhance our understanding of anatomical systems. We all recall the experience: donning used safety goggles, a white lab coat, and gloves (which are essential, of course). A few more pieces of equipment come to mind; what about microscopes? Clear lab test tubes and cylinders?

How about the atmosphere?

Do you remember the classic, large, laminated human body anatomy poster that hangs in most science classes? Lastly, we should also recall the iconic periodic table poster!

This educational approach primarily takes place within a group of teams or classmates. Classmates are encouraged to collaborate and share responsibilities through participation while fostering and learning positive social and multicultural etiquette. The teams will comprise participants from diverse backgrounds with varying ideologies, skills, and interests (Thomas, 2002) to address and achieve a common objective.

In 2002, collaborative efforts were undertaken to address and achieve a common objective, mission, or goal.

Let us discuss leaf litter. It is pertinent to communicate the significance of leaves.

Through photosynthesis, leaves absorb carbon dioxide and water, utilizing solar energy to convert these elements into chemical compounds, principally sugars, which sustain the tree. Notably, the tree also produces and releases oxygen as a by-product of this chemical reaction (Stancil, 2015). During the colder autumn and winter months, particularly in urban areas and local neighbourhoods, leaf litter can be observed due to fallen leaves on sidewalks and driveways. This accumulation often obstructs municipal catch basins, thereby hindering water flow during periods of heavy rainfall..

Figure 1.1 © Moses Filter Drain Team, 2025. *Fallen leaves on sidewalks and driveways. This accumulation often obstructs municipal catch basins, hindering water flow during periods of heavy rainfall.*

In the fall, water is compelled to pass through the obstructed debris, predominantly leaves, which may lead to the degradation of organic material. This process releases pollutants and floating debris attached to the leaves via municipal catch basins (Pitt and Field, 2004).

As delineated in Field (2004), the initiative to reduce clutter aims to mitigate the prevalence of litter in our local streams, rivers, lakes, and oceans. These water bodies harbour harmful microbes that often remain undetected, concealed, and camouflaged.

Figure 1.2 © Moses Filter Drain Team, 2024. Reduced urban runoff is the most common, and we aim to focus on Leaf Litter & micro-particle waste. [We suggest viewing the photo in colour.]. British Columbia, Canada

The above photograph illustrates (figure 1.2 above and 1.3 below) the leaf litter collected in one of the three filter cleanouts in November (during the fall) as part of the filter cleanouts to perform the data analysis process. Notably, small to medium-sized debris was carefully separated from the leaves before an extensive examination of the potential for leaf litter to harbour micro- and nano-plastic waste washed into municipal catch basins in urban metropolitan areas.

This illustration aimed to investigate the impact of leaf litter traveling (rain) downstream on cement or concrete (roads) towards municipal and private property grates-catch basins, potentially carrying billions of micro- and nano-plastics to our local marine ecosystem as a significant transport vehicle, hidden and undetected. The idea is that the breakdown of leaf litter proceeds in three distinct temporal stages: leaching, conditioning, and fragmentation (Gessner et al., 1999). The leaching stage can help understand "pollutants to describe portions of the litter breakdown process in freshwaters" (Gessner, 1999), such as street salt sprinkled on city streets during the winter and snowy season months.

Figure 1.3: © Moses Filter Drain Team, 2024. *"Leaf litter is a collection of fallen leaves, twigs, seeds, flowers, and other woody debris."*

Leaf litter can be described as a collection of fallen leaves, twigs, seeds, flowers, and other woody debris that naturally accumulates on the ground in forest environments, all of which subsequently enter the decomposition pathway in terrestrial ecosystems (Johnson & Catley, 2002; Berg & McClaugherty, 2008). However, our data analysis focused on leaf litter within the metropolitan urban infrastructure because of its adhesive ability to capture micro-particle waste (debris). "The problems are unsightly when not maintained," "cost the city excessively," and "drip sap or sticky residue on parked cars" (on concrete pavements) (Lohr et al., 2004). We propose that because of leaf litter's camouflaging abilities, "people in metropolitan areas" have "recognized that there are problems associated with" fallen leaves and "trees" and that "the issues arise from" an accumulation of leaves "on sidewalks," along with their costs to cities" (Lohr et al., 2004). Furthermore, it negatively floods our local streams, rivers, lakes, and oceans with micro- and nano-plastic waste in astronomical quantities (Lohr et al., 2004).

Figure 1.4: © Moses Filter Drain Team, 2024. [We suggest viewing the photo in colour.]

The photo (figure 1.4) illustrates leaf litter harbouring several unidentifiable micro-particle wastes; visually, we can observe non-organic particles attached to the leaves. Since the picture was taken during winter, in addition to other harmful physical particles, we also observe street salt, camouflaged and passing through municipal and private catch basin grates, accumulating and ultimately increasing sodium levels in gardens and local streams.

"Both dissolved organic carbon (DOC) and salt concentrations are rising in northern hemisphere watersheds, yet there has been little investigation of sodium as a potential mechanism for DOC increases" (Steele, 2013)."Dead plant material (leaf litter) and animal residue are gradually decomposed until their original identity is no longer recognizable, at which point they are considered soil organic matter (SOM)." (Chapin et al., 2002). Analyzing and observing the natural process of decomposing leaves, we observed particle-attached to the leaves well-camouflaged pollutants in urban runoff, including tired thread pollution (rubber) litter and heavy metal; our focus shifted to micro-nano plastics interacting with leaf litter.

What Are Microplastics?

"Microplastics form from the degradation of larger plastic pieces or are intentionally produced as materials of this small size. They are often defined as particles greater than 5 mm, although they frequently range to several micrometres" (Cole et al., 2011). Additionally, intentionally manufactured primary microplastics include microbeads found in personal care products and industrial abrasives. Microbead agents, coatings, paints, resins, and pellets are used to manufacture finished plastic products (Hale, 2020).

What Are Nanoparticles?

"Nanoparticles are ultra-small particles with properties," and "nanoparticles and nanomaterials may exhibit harmful properties when leaked" into the "environment" and eventually washed down municipal catch basins during heavy rainfall, leading to pollution within our marine ecosystem (Kumari et al., 2022). Is it shocking that "materials of 300 nm or less in diameter" of nanoparticle pollution "can be taken up by individual human cells"? (Garnett and Kallinteri, 2006).

We shall examine the interaction of micro and nano glass with various environmental communities. A significant factor contributing to the toxicity of glass is its ingestion by marine organisms and subsequent transfer into the food web. Such pollution may lead to numerous detrimental consequences. Furthermore, micro- and nanosized contaminants pose a more significant threat than macro-pollutants due to their diminutive size (Kumari et al., 2022). The Pacific Northwest region, specifically Metro Vancouver, Canada, receives an average annual rainfall of approximately 1200 to 1600 millimetres (Vancouver Weather Stats, 2023).

Quantitative analysis: We created a rectangular area measuring 60 x 30 mm^2 at approximately 70%.

During intense rainfall, leaf litter and micro- and nanoparticles are clandestinely transported using custom-made camouflage vehicles. These materials are surreptitiously funnelled into urban drainage systems and ultimately discharged into our local water sources. It has been noted that "the small-size particles can easily intermingle with the environment," which may facilitate their translocation through our food and water cycles (Kumari et al., 2022). Interactions involving terrestrial organisms and micro- and nano-glass, as well as with marine species, are significant. Furthermore, the interactions between soil and micro- and nano-glass and the dynamics involving plant-microbe-micro/nano-glass interactions are particularly relevant. Additionally, the interactions involving domestic pets, such as cats and dogs, during outdoor activities also warrant attention (Kumari et al., 2022).

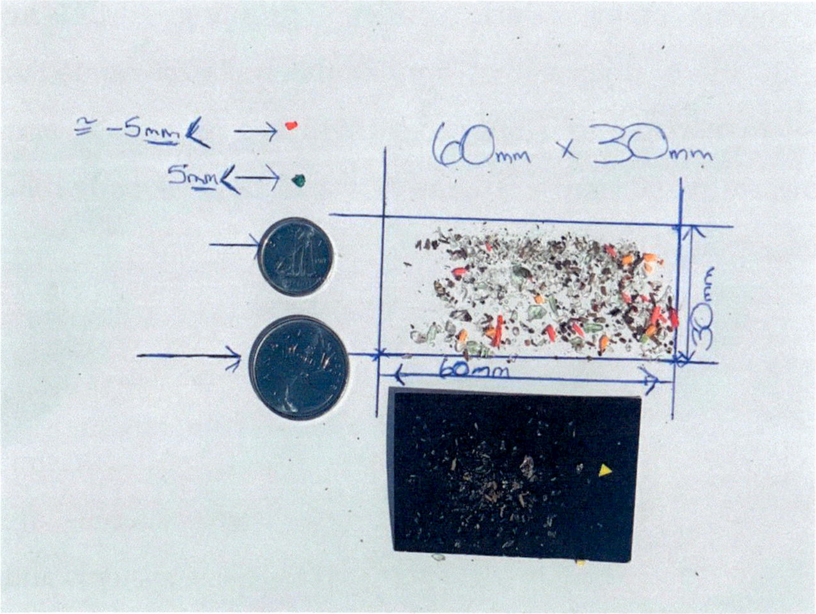

Figure 1.5: © Moses Filter Drain Team, 2024. [We suggest viewing the photo in colour.]

Our study used a microscope to quantify the number of pollution particles in a defined rectangular area. The black background rectangle was designed to showcase the crystalline appearance of particles derived from a mixture of plastics and shattered glass, indicating the density of pollution pieces per rectangle. It was observed that approximately 16% of glass drinking containers contribute to 53% of glass beverage bottles, car windows, and automotive headlights, which is attributed to incidents involving motor vehicle accidents (Whiting, 1998). Our estimations suggest that approximately 18,750 to 19,999 micro/nanoparticles are present within a rectangular area measuring 60 mm^2 x 30 mm^2 between the black and white background rectangles.

Figure 1.6: © Moses Filter Drain Team, 2024. [We suggest viewing the photo in colour.]

During the 6-month check-in for our catch basin pilot with the Green Infrastructure Department, we utilized eight different GI catch basin locations. Figure 1.6 (above) visually depicts the micro/nanoparticle waste that was captured and analyzed. The base box measures 640 mm in diameter and 230 mm^2. We estimate that using a box with dimensions of 60x30 mm, compared to the larger one, accounts for roughly 0.90 - 2% of the total. Therefore, we can confidently state that we have prevented over 1.1 billion micro/nanoparticles from entering our municipal ocean catch basins in six months via the eight municipal catch basin locations.

The camouflaging and adhesive properties associated with the presence of microplastics (debris) embedded within the viscous, discernible attributes of decomposing leaf litter—characterized by its dark coloration and mushy texture—render it an effective medium for the adhesive capture of microplastic particles as they are conveyed through urban drainage systems during periods of substantial rainfall (refer to Figure 1.6 below). A significant mass loss is observed during the decomposition phase of leaves; approximately 67% of this loss is often exacerbated by the addition of litter. Throughout our analytical sessions, the observed mass loss facilitated the extraction and segregation of microplastic particles adhering to decomposing leaves or leaf litter. Furthermore, the mass loss of leaves was further augmented when litter mixtures were introduced (Gartner & Cardon, 2004). This finding prompted a reevaluation of our research focus, leading us to closely examine leaf litter as a vector for the transportation of microplastics into our largely undetected marine ecosystems.

Figure 1.7 - © Drain Team, 2024. [We recommend viewing the photo in colour]

The image above depicts a catch basin during heavy rainfall. At this time, pollutants are clearly transported along urban streets and directed through the basin grates. Leaf litter's concealing effects, which obscure and intermittently release a considerable volume of microplastic waste into municipal drainage systems, go unnoticed while simultaneously retaining debris from urban runoff.

Waste may be held together by an accumulation of leaves that grows during the fall and winter seasons. These leaves are then raked or washed down city drains, harming our ecosystems. Our findings connect decomposing leaves to the transport of small litter (floatables), which negatively impacts our water, marine habitats, and humans.

Figure 1.8 © Moses Filter Drain Team, 2025. Our findings link decomposing leaves to the movement of small litter (floatables), which adversely affects our water, marine habitats, and humans.

It may be surprising to discover that nearly 100% of the debris, commonly referred to as floatable pollutants, enters municipal catch basins due to rainfall and wind, subsequently flowing into our local marine ecosystem. A comprehensive database is currently being developed from the data collected during the scheduled cleaning of catch basin filters. This data, which encompasses observations and analyses involving approximately 150 types of litter and debris per basin (Grey et al. 2007), identifies microplastics and particle waste drifting down our municipal drains as the most detrimental to marine life (refer to figure 1.6 above). This situation arises because urban rainwater runoff constitutes a significant source of pollution that is discharged through sewer and drainage systems into receiving waters (Thomson et al., 2019).

During data analysis, we categorize debris and litter into distinct classifications before quantifying and entering the data into our designated platform. While segregating and organizing the debris, the solid materials were relatively straightforward to identify and quantify. The small—to medium-sized debris comprised solid plastics, films, and single-use items, including utensils and public transportation tickets or passes. Initially, we weighed the collected leaves to furnish quantitative data regarding the organic matter represented by these leaves.

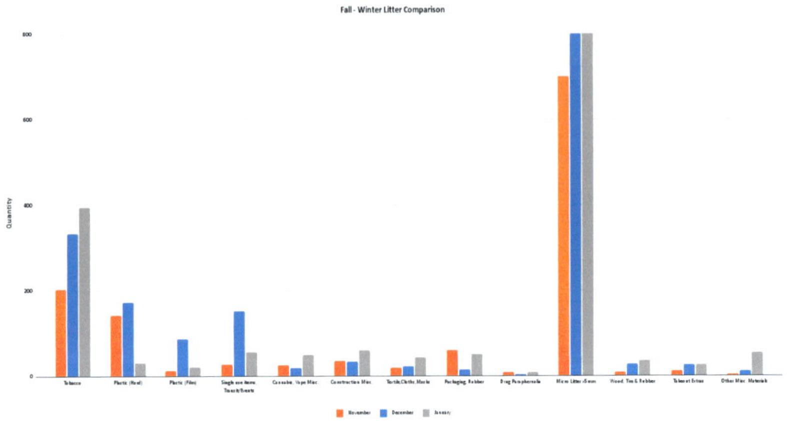

Graph 1.0: Fall- Winter (litter) Comparison: November (orange), December (blue), January (grey)

The estimated number of micro- and nanoparticle pollutants was significantly reduced within our Moses Filter Data Entry Platform, allowing us to observe the other debris we collected and analyzed from washing down municipal catch basins. However, we observed a spike in micro- and nanoparticle plastic waste in more significant quantities during these three months, likely due to the abundant falling leaves in autumn and winter (Q4 of the year). During the heavy rainfall season, city drains become clogged with leaves; we hypothesize that this is when the massive migration begins.

Based on this observation, it can be estimated that the percentage of microplastics in sticky leaf litter in urban areas is likely to increase during the autumn and winter months due to city drains becoming clogged with leaves, resulting in a migration of microplastics during heavy rainfall. Tobacco ranked second (2nd), while other materials such as leaves, sand, rocks, mud, and sediment ranked third (3rd), and single-use items placed fourth (4th) out of the top 13.

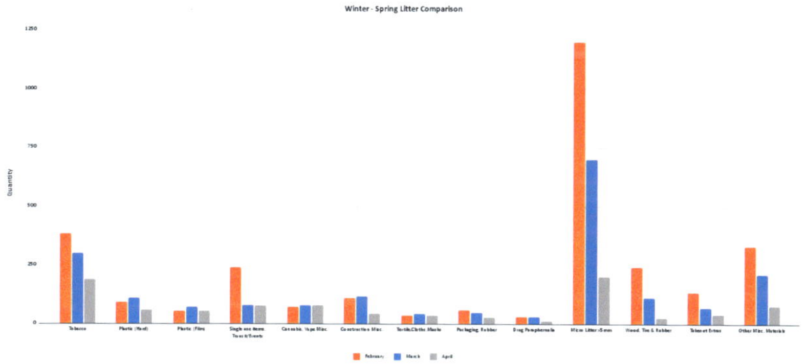

Graph 2.0: Winter–Spring (litter) Comparison: February (orange), March (blue), and April (grey)

The estimated number of micro- and nanoparticle pollutants was drastically minimized within our Moses Filter Data Entry Platform to factor in the other collected debris we were reducing from washing down municipal catch basins and the ecosystem. Tobacco ranked second (2nd), different materials (leaves, sand, rocks, and mud, to name a few) ranked third (3rd), and single-use items ranked fourth (4th) out of 13.

Unlike larger fragments, microplastics are not easily visible to the naked eye; even resin pellets (microplastics) mixed with sand are hard to discern. Net sampling fails to collect the smaller microplastics, and no acceptable standard procedure exists for their enumeration in water or sand. The following is a suggested procedure derived from published reports and the author's personal experience (Andrady, 2011).

Concern arose over potential negative impacts on marine life. While plastic waste in the environment presents a global and multifaceted issue, media and scientific focus has primarily been "ocean-centric" (Hale, 2020). Furthermore, this encompasses "the fate, properties, and distribution of microplastics (MPs) and nanoplastics (NPs) in soil." As a result, "most of the 300 million tonnes of plastics produced each year end up in the environment, with soil serving as a long-term sink for this plastic debris" (Pathan, 2020).

Figure 1.9: © Moses Filter Drain Team, 2025. Microscope view of reduced microparticles entering our local marine ecosystem. *"300 million tonnes of plastics produced yearly end up in the environment, with soil serving as a long-term sink for this plastic debris"*

Micro-Rubber & Leaf Litter

Approximately 3 billion new tires are manufactured each year, yet a troubling 800 million are discarded annually. Pollutants released from tire (rubber) wear significantly threaten the environment and human health. When tires deteriorate, they emit pollutants into the air, water, and soil, extensively affecting various ecosystems (Mayer, 2024).

Figure 1.10: © Moses Filter Drain Team, 2024. "Rubber particles from tire wear are effectively camouflaged, seamlessly blending with fallen leaves and small – to medium-sized sticks".

However, our Moses Filter Data indicated that rubber particles from tire wear are effectively camouflaged, seamlessly blending with fallen leaves and small—to medium-sized sticks. These visibly conceal the harmful pollutants from tire wear. Figure 1.6 illustrates the middle-to micro-sized rubber pieces under a microscope that seemingly blend in with natural tree branch pieces.

Figure 1.11: © Moses Filter Drain Team, 2024. [We suggest viewing the photo in colour.]

There are various ways to quantify physical tire wear (pieces), which come in multiple sizes, dimensions, strands, and shapes. Ecologist Paul Mayer (2024) found that a single tire can shed between two and fourteen pounds of rubber particles due to road wear (from initial use to disposal). These particles may be small enough to be picked up by the wind and carried for up to a month before being deposited on land (Figure 1.6).

The research found that one tire will shed between two and fourteen pounds of rubber particles due to road wear (from initial use to initial disposal). These particles may be small enough to be picked up by the wind and carried for up to a month before being deposited on land. Larger particles can be caught in stormwater runoff and transported along curbs and through stormwater systems, where they are typically deposited into a local waterway. The constituents of these particles—pollutants such as microplastics, heavy metals, hydrocarbons, and other toxic chemicals—can then contaminate local water and soil.

Zogg (2019) compared rubber to another polymer that pollutes our air, water, and organisms: fine particles from tire abrasion. These particles enter our soil and air through the road surface or are washed away by artificial turf.

Empa researchers have calculated that over the last 30 years, from 1988 to 2018, approximately 200,000 tonnes of micro rubber have accumulated in the environment in Switzerland.

Figure 1.12: © Moses Filter Drain Team, 2025. Microscope view of reduced microparticles entering our local marine ecosystem. "These fine particles from tire abrasion enter our soil and air through the road surface or are washed away" (Zogg 2019).

The amount of micro rubber entering our environment through municipal catch basins and leaf runoff is significant and highly impactful. It seamlessly blends with some of the most harmful urban runoff into our local marine ecosystem—the trickle-down effects are unknown. The combination of fallen leaves and their concealing properties significantly impacts the ecosystem, raising growing concerns and dangers.

Figure 1.13 © Moses Filter Drain Team, 2025. *Microrubber entering our environment through municipal leaf runoff is significant, seemingly blending in and camouflaging with leaves and other unintended runoffs.*

In conclusion, leaf litter, which can possess camouflage characteristics, has a striking effect on the environment by creating an optical illusion of natural decomposition and hindering the detection and removal of harmful waste and micro/nano plastics from city catch basins during heavy rainfall seasons before they reach the local marine ecosystem.

Figure 1.14 © Moses Filter Drain Team, 2025. A discharge location within a local community is where groundwater is expelled or discharged onto the land's surface. *The significant results can have serious consequences; as such, waste and pollution accumulation can negatively*

impact soil quality, water quality, and ecosystem health (Spiralytics, 2023), as illustrated in Figure 1.14 above.

The significant results can have serious consequences. Waste and pollution accumulation can negatively impact soil quality, water quality, and ecosystem health, burdening municipal infrastructure. We aim to inspire by highlighting critical future research areas for engineers, environmental scientists, policymakers, and community members because we share the same drinking water.

References

Andrady, Anthony L. "Microplastics in the Marine Environment." *Marine Pollution Bulletin*, vol. 62, no. 8, Aug. 2011, pp. 1596-1605.

Barnes, David K. A., François Galgani, Richard C. Thompson, and Morton Barlaz. "Accumulation and Fragmentation of Plastic Debris in Global Environments." *Philosophical Transactions of the Royal Society B: Biological Sciences*, July 2009, https://doi.org/10.1098/rstb.2008.0205. Accessed 12 May 2023.

Browne, M. A., T. Galloway, and R. Thompson. "Microplastic – An Emerging Contaminant of Potential Concern?" *Integrated Environmental Assessment and Management*, vol. 3, 2007, pp. 559-561.

Chalmers, Alan F. *What Is This Thing Called Science?* 4th ed., 2013.

Chapin, F. Stuart, Pamela A. Matson, and Harold A. Mooney. *Principles of Terrestrial Ecosystem Ecology: Terrestrial*

Production Processes. Springer, New York, NY, https://doi.org/10.1007/0-387-21663-4_6, pp. 151-175.

Chiyangwa, Tawanda Blessing, and P. M. (Trish) Alexander. "Rapidly Co-Evolving Technology Adoption and Diffusion Models." *School of Computing, University of South Africa, Science Campus*, Florida Glen, Johannesburg, South Africa, June 2015.

Coast Mental Health. *Landscaping With Hearts.* Vancouver, British Columbia, Canada, Coast Mental Health Foundation, https://www.coastmentalhealth.com/landscaping-with-heart/.

Cole, M., Lindeque, P., Halsband, C., and Galloway, T. S. "Microplastics as Contaminants in the Marine Environment: A Review." *Marine Pollution Bulletin*, vol. 62, 2011, pp. 2588–2597, https://doi.org/10.1016/j.marpolbul.2011.09.025. Accessed 10 May 2023.

Garnett, P., and K. Kallinteri. "Nanomedicines and Nanotoxicology: Some Physiological Principles." *Occupational Medicine*, vol. 56, no. 5, 2006, pp. 307-311, https://doi.org/10.1093/occmed/kql052.

Gartner, Tracy B., and Zoe G. Cardon. "Decomposition Dynamics in Mixed-Species Leaf Litter." *Oikos*, 16 Jan. 2004, https://doi.org/10.1111/j.0030-1299.2004.12738.x.

Gessner, Mark O., Eric Chauvet, and Mike Dobson. "A Perspective on Leaf Litter Breakdown in Streams." *Oikos*, vol. 85, no. 2, May 1999, pp. 377-384, https://www.jstor.org/stable/3546505. Accessed 10 Apr. 2023.

Graham, E. R., and J. T. Thompson. "Deposit- and Suspension-Feeding Sea Cucumbers (Echinodermata) Ingest Plastic Fragments." *Journal of Experimental Marine Biology and Ecology*, vol. 368, 2009, pp. 22-29.

Grey, M. Gary, and Frank Oliveri. "The Role of Catch Basins in a CSO Floatables Control Program." *New York City Department of Environmental Protection, Sewer Analysis Bureau of Water Supply & Wastewater*, 2007.

Hale, R. C., Seeley, M. E., La Guardia, M. J., Mai, L., and Zeng, E. Y. "A Global Perspective on Microplastics." *Journal of Geophysical Research: Oceans*, 2020, https://doi.org/10.1029/2018JC014719. Accessed Apr. 2023.

Johnson, E. A., and K. M. Catley. *Life in the Leaf Litter.* Center for Biodiversity and Conservation, American Museum of Natural History, 2002.

Kumari, Sonu, Swati Agarwal, and Suphiya Khan. "Micro/Nano Glass Pollution as an Emerging Pollutant Shortly." *Journal of Hazardous Materials Advances,* vol. 6, 2022, p. 100063, https://doi.org/10.1016/j.hazadv.2022.100063. Accessed 5 Apr. 2023.

Lohr, VI, Pearson-Mims, CH, Tarnai, J, and Dillman, DA. "How Urban Residents Rate and Rank the Benefits and Problems Associated with Trees in Cities." *Journal of Arboriculture,* vol. 30, no. 1, 2004, pp. 28-35.

Molnar, Charles, et al. *Concepts of Biology - 1st Canadian Edition.* 2013, https://rb.gy/naios. Accessed 10 May 2023.

Moses Filter Inc., and Drain Team Founder/CEO Jean-Paul Kamand. Vancouver, BC, Canada.

Partridge, J, "Conducting a Science Investigation in a Primary Classroom." *Teaching Science,* vol. 52, no. 2, Winter 2006, pp. 44-45, https://rb.gy/tf23t. Accessed 18 May 2023.

Pascoal, Cláudia, and Fernanda Cássio. "Contribution of Fungi and Bacteria to Leaf Litter Decomposition in a

Polluted River." *Journal of Aquatic Microbial Ecology*, 1 Sept. 2004, https://rb.gy/eu6d7. Accessed Apr. 2023.

Pathan, S. I., et al. "Soil Pollution from Micro- and Nanoplastic Debris: A Hidden and Unknown Biohazard." *Sustainability*, vol. 12, 2020, p. 7255, https://doi.org/10.3390/su12187255. Accessed 1 May 2023.

Pitt, Robert, and Richard Field. "Catch Basins and Inserts for the Control of Gross Solids and Conventional Stormwater Pollutants." *World Water and Environmental Resources Congress*, 2004.

Ryan, P. G., et al.. "Monitoring the Abundance of Plastic Debris in the Marine Environment." *Philosophical Transactions of the Royal Society B: Biological Sciences*, vol. 364, 2009, pp. 1999-2012, https://doi.org/10.1098/rstb.2008.0207.

Spiralytics. *A Homeowner's Guide to P...* Published 30 May 2023. https://shorturl.at/lLtiU. Accessed 25 Feb. 2025.

Stancil, Joanna Mounce. "The Power of One Tree - The Very Air We Breathe." *U.S. Forest Service in Forestry*, 17 Mar. 2015, https://l8.nu/rwtN. Accessed 20 Mar. 2023.

Steele, M. K., and J. A. Aitkenhead-Peterson. "Salt Impacts Organic Carbon and Nitrogen Leaching from Senesced Vegetation." *Biogeochemistry*, vol. 112, 2013, pp. 245–259, https://rb.gy/tq5h4. Accessed 15 Apr. 2023.

Thomas, G. *Effective Classroom Teamwork: Support or Intrusion?* 1st ed., Routledge, 1992, https://doi.org/10.4324/9780203038147.

Thomson, Matthew, et al. *Rain City Strategy: A Green Rainwater Infrastructure and Rainwater Management Initiative.* The City of Vancouver, 5 Nov. 2019, https://vancouver.ca/files/cov/rain-city-strategy.pdf. Accessed 16 Apr. 2023.

Zogg, Cornelia. "Rubber in the Environment: Where the Tread from Our Tires 'Disappears' To." *Swiss Federal Laboratories for Materials Science and Technology*, 14 Nov. 2019, https://phys.org/news/2019-11-rubber-environment.html. Accessed 19 Nov. 2024.

Acknowledgement

On behalf of the Moses Filter Drain Team, we would like to express our gratitude for the support from the City of Vancouver's Green Infrastructure Engineering Department, Vancouver Lower Mainland (BC), and Coast Mental Health.

The City of Vancouver's Green Infrastructure Engineering Department provided invaluable insights, identified drainage locations, and offered ongoing support throughout our pilot. The department empowered us to move forward in our mission from our invaluable experience of reducing unintentional urban runoff from entering the Pacific Ocean, the Fraser River, and other local marine bodies.

We are equally grateful to Coast Mental Health (Landscape with Hearts) for contributing its expertise in the mental health aspects of urban ecosystems through customized employment. Additionally, the incredible and talented creators of the Moses Filter Data Entry Platform have been essential for our research data analysis.

To my brother Raoul Kibangula, thank you for your support and assistance in making this a reality. The combined efforts of these organizations and individuals have been instrumental to our success.

We felt compelled to publish this comprehensive report as a book. It encourages conversations and awareness about climate adaptation and provides a detailed analysis of the intersection between unintentional urban runoff and sustainability practices. This information is significant for understanding and implementing sustainable innovations.

Book Report Overview:

This case study investigates the essential function of leaf litter in transporting microplastics and various pollutants into municipal catch basins, thereby impacting marine ecosystems. Using the advanced Moses Filter filtration system, the research offers invaluable insights into urban runoff during intense rainfall, illustrating how leaf litter acts as a significant but frequently neglected vehicle for harmful microplastics.

Key Findings:

The investigation highlights the connection between leaf litter and the transport of microplastics, particularly during heavy rainfall, fall, and windy seasons. We emphasize the environmental impacts of leaf litter in the fall season.

- The collected data indicates that leaf litter can inadvertently discharge pollutants into local waterways, underscoring the urgent need for action to mitigate this issue.

Target Audience:

This report is intended for a broad audience, including environmental scientists, municipal planners, educators, and anyone concerned about clean water and microplastics in our marine ecosystem. It aims to raise awareness of the often-overlooked impact of leaf litter and the importance of innovative solutions. We encourage innovators to share their ideas, as we are always open to presenting potential solutions to these challenges.

About the Author:

Jean-Paul (JP) Kamand is an accomplished entrepreneur committed to achieving business success while making a meaningful impact on the world. His work with Moses Filter demonstrates his dedication to social responsibility and environmental sustainability while addressing the urgent need for customized employment for *Individuals With All Abilities* (IWAA). Based in beautiful British Columbia, Jean-Paul is passionate about positively contributing to his community and extending that influence beyond.

For personal guest speaker opportunities, comments, or bulk book orders, please contact Jean-Paul at…

E-mail: jpkamand@gmail.com

Made in the USA
Monee, IL
05 November 2026

67379440R00033